THE ADVENTURES OF LUNA THE ASTRONAUT

First edition. December 18, 2024.

ISBN: 979-8224155385

Written by Anila Asif.

"The Adventures of Luna The Astronaut"

Chapter 1:
Luna's Dream of the Stars

Luna had always been fascinated by the stars. Every night, she would climb into bed, pull the covers up to her chin, and stare out of her window at the vast, twinkling sky. She loved how the stars seemed to wink at her, as if they were sharing secrets of the universe. Luna often dreamed of exploring the farthest reaches of space, flying on rockets, visiting distant planets, and maybe even meeting friendly aliens.

She imagined herself floating in space, weightless, as she gazed down at the Earth far below. But most of all, Luna dreamed of one day traveling to the International Space Station (ISS), the incredible space station that orbited high above the Earth. She had learned about it in school and had become captivated by the idea that astronauts lived and worked there, studying space and conducting experiments.

It seemed like the ultimate adventure, and Luna knew that if she ever got the chance, she would jump at it.

One evening, after finishing her homework, Luna ran outside to her backyard, where the stars were especially bright. As she lay on the soft grass, she made a wish to travel to space. She closed her eyes tightly and whispered to the stars, "I wish I could go to the space station and see everything with my own eyes."

As Luna gazed up at the night sky, something amazing happened. The stars seemed to twinkle even brighter, and she thought she saw something unusual—a small, glowing object falling gently from the sky. It landed softly in front of her, right in the middle of the backyard! Luna sat up, wide-eyed, and reached out to pick it up.

To her amazement, it was a letter. The envelope was decorated with pictures of rockets and stars.

Luna carefully opened the letter, and inside, she found a message that made her heart race with excitement:

"Dear Luna,

We have received your wish, and we are pleased to invite you on a very special mission. You have been selected to visit the International Space Station. Prepare for an adventure that will take you beyond your wildest dreams!

Sincerely,

The Space Team"

Luna's eyes went wide with disbelief. Could this be real? She read the letter again and again, her excitement growing with every word. Her dream was about to come true!

Fun Fact: The International Space Station (ISS) orbits Earth about 250 miles above the surface, and astronauts live there for months at a time to study space, conduct experiments, and learn more about how humans can live in space. The ISS travels at a speed of around 17,500 miles per hour, which means it orbits the Earth about 16 times a day!

Luna couldn't wait to tell her parents about the incredible news. She jumped up, feeling like she was already floating in space. Little did she know, her incredible adventure was just beginning!

Chapter 2:
The Rocket Ride

The day Luna had been waiting for had finally arrived. She stood at the launchpad, her heart pounding with excitement. She could hardly believe it—she was about to board a real space shuttle and fly to the International Space Station! Around her, astronauts in their bright white space suits were checking equipment, making final preparations, and ensuring everything was ready for the incredible journey ahead.

Luna was handed her own space suit, and as she slipped it on, she felt like a true astronaut. She couldn’t stop smiling as she imagined what it would be like to float in zero gravity and look down on the Earth from above. "This is it," she whispered to herself. "I’m going to space!"

The shuttle’s doors closed, and Luna took her seat. She felt a little nervous, but the excitement quickly took over.

The countdown began, and Luna gripped the armrests of her seat as she listened to the sound of the engines starting. The countdown reached zero, and with a roar, the rocket engines fired up.

"Lift-off!" shouted the mission commander.

Suddenly, the entire shuttle shook as it launched into the sky. Luna's stomach did a little flip as the rocket soared higher and higher, breaking free from Earth's gravity.

The ground below quickly shrank into a small dot, and soon, all that could be seen were the clouds and the endless blue sky.

As they ascended through the atmosphere, Luna looked out the window. She could see the Earth growing smaller and smaller, its oceans, continents, and cities disappearing beneath a blanket of clouds.

It was hard to believe that just a few minutes ago, she had been standing on the ground. Now she was floating in space, traveling faster than she had ever imagined.

The shuttle continued on its journey, and soon, Luna saw something incredible—the Moon! It was bigger than she had ever imagined, glowing a soft gray against the blackness of space. Luna pressed her face to the window, gazing at the craters and valleys that covered its surface. She had seen pictures of the Moon before, but nothing compared to the real thing.

“Look at those craters!” Luna exclaimed. “How did they get there?”

The astronaut sitting next to her, a friendly woman named Commander Sarah, smiled. “Great question, Luna! The Moon’s craters were created millions of years ago when big asteroids or comets collided with it. These space rocks hit the Moon with incredible force, leaving behind huge holes in the surface.”

Luna nodded, fascinated. “So, the Moon wasn’t always like this?”

Commander Sarah shook her head. “No, Luna. The Moon was formed when a giant object collided with Earth a long time ago. The impact sent pieces of Earth flying into space, and over time, these pieces came together to form the Moon.” Luna could hardly believe it. She had always wondered about the Moon, but now she was seeing it up close, understanding how it was shaped by space collisions.

As the shuttle flew closer, she could see the sharp edges of the craters, some of them so deep that they looked like dark shadows on the Moon's surface.

Fun Fact: The Moon was created millions of years ago when a giant object collided with Earth. The debris from the collision flew into space, and over time, it clumped together to form the Moon! That's why the Moon has so many craters —because space rocks have been hitting it for billions of years.

Even today, small meteoroids crash into the Moon, but it doesn't have an atmosphere like Earth to burn them up. The shuttle zoomed past the Moon, and Luna could see the entire Earth below her now—a beautiful blue and green ball, spinning slowly in the blackness of space. She felt so small, yet so connected to everything.

The vastness of space stretched out before her, and Luna knew that this was just the beginning of her incredible adventure.

As the rocket continued on its path to the International Space Station, Luna gazed out at the stars, wondering what other amazing things she would see on her journey.

Chapter 3: Floating in Space

Luna saw something incredible—the Moon! It was bigger than she had ever imagined, glowing a soft gray against the blackness of space. Luna pressed her face to the window, gazing at the craters and valleys that covered its surface. She had seen pictures of the Moon before,

Luna felt a gentle push against the floor, and before she knew it, she was drifting up into the air. She gasped and laughed, her feet leaving the ground. “This is incredible!” she said as she floated effortlessly. It was like being in a dream, where nothing was heavy and everything could fly.

She reached out to grab a pencil that was floating in front of her, but it zoomed past her hand, slipping away as if it had a mind of its own.

Luna tried again, but this time, her hand touched the pencil, and it bounced away gently. She burst into laughter, trying to catch it again. It was harder than she thought!

A friendly astronaut, Captain Liam, floated by and waved. “Welcome to space, Luna! This is what we call ‘microgravity.’ It means that the force of gravity is much weaker here than on Earth, so things don’t fall to the ground like they do back home.”

Luna watched as Captain Liam tossed a water bottle into the air. It didn't fall; instead, it floated gently in the middle of the room. Luna reached out and touched it. It felt strange. She could push the bottle around with just a light tap of her finger, and it would float in any direction she sent it.

"So, in space, nothing falls?" Luna asked, still amazed.

"That's right!" Captain Liam replied with a smile. "In space, we don't feel the pull of gravity the way we do on Earth.

Things float because there's nothing pulling them down. That's why we have to be careful not to let things float away and get lost!"

Luna thought about the Earth for a moment. On Earth, gravity kept everything in place. When she dropped her pencil at school, it always fell to the ground. But here, in space, everything was weightless, like a balloon drifting gently in the air.

She pushed herself off a wall, and, to her surprise, she floated across the room, spinning slowly in the air. “This is so much fun!” Luna giggled, tumbling gently through the space station.

She had to hold on to the walls or any object nearby to stay in one place. Otherwise, she would float away. Luna tried to grab a floating book, but as she reached out, it spun away just out of her grasp. She laughed even harder, trying to catch it and giggling as she tumbled through the room like an astronaut in training.

The sensation of floating was amazing, and Luna realized that space wasn't just an adventure to look at—it was something to experience with every part of her body. She felt light, free, and like she could soar through the air forever.

Fun Fact: In space, astronauts experience "microgravity," which makes them float. This happens because gravity is weaker up there than on Earth. Microgravity occurs because the space station is constantly falling toward Earth, but it is also moving so fast forward that it keeps missing.

This creates the sensation of weightlessness, allowing everything—people, pencils, water bottles—to float! Luna felt like she was flying through the air like a bird. She spun around in a slow circle, her feet no longer touching the floor. For the first time, she truly understood what it felt like to be free from gravity. In space, she could float, glide, and twist in the air without any effort. It was like a never-ending game of flying and floating—and Luna loved every moment of it!

As she floated past the windows of the space station, she looked out at the Earth below. The beautiful blue planet seemed so far away now, but in a way, it felt closer than ever. Luna realized that she was part of something much bigger than herself—a journey that spanned the vastness of space.

She couldn’t wait to see what other wonders awaited her in this amazing place!

Chapter 4:
A Tour of the Space Station

After Luna had spent some time floating around and getting used to the weightlessness of space, it was time for her to explore the rest of the space station. Commander Sarah greeted her with a smile and gave her a tour of the station, showing her all the amazing things that made life in space possible. As Luna floated behind her, she looked around in awe.

The space station was so much bigger than she had imagined! It was like a huge metal house, filled with rooms and hallways, each designed for a specific purpose. The walls were lined with controls, computers, and strange-looking equipment. There were also small windows everywhere, offering breathtaking views of space. "This is our command center," Commander Sarah said, pointing to a room filled with screens and buttons. "This is where we monitor everything happening inside and outside the station.

We make sure that everything is working properly so we can keep learning about space."

Luna's eyes grew wide. "It's like a big spaceship!"

"Exactly!" Commander Sarah replied. "But it's also much more than that. The International Space Station is like a giant science lab. Astronauts live and work here, studying space, conducting experiments, and learning how humans can survive in space for long periods of time."

As they continued their tour, Luna noticed several rooms filled with strange-looking machines and tools. “What are all these things for?” she asked. “These are some of the experiments we’re working on,” Commander Sarah explained. “In space, we can do experiments that aren’t possible on Earth. We can study how plants grow without gravity, how our muscles and bones change in space, and how materials behave in microgravity.

On Earth, plants grow toward the light, but in space, they have to adapt. We're studying how they grow here so we can use them to help astronauts grow food during long missions, like trips to Mars."

Luna couldn't believe it. "So, astronauts might grow their own food on Mars someday?"

"That's the idea!" Commander Sarah replied. "Space farming could be a big part of future missions to other planets. If we can grow food in space, it will be easier for astronauts to live there for long periods."

They moved on to another part of the station, where Luna saw astronauts floating as they worked with computers, checking data, and making sure everything was running smoothly. “How do astronauts sleep in space?” Luna asked, curious.

“Great question!” Commander Sarah said with a smile. “Astronauts sleep in small sleeping pods, like little rooms that float. We have to strap ourselves into sleeping bags to stay in one place, or else we’d float around all night!”

Luna imagined floating while she slept and giggled. It seemed so different from her bed at home, where she could roll over and snuggle under the covers. Here, there was no "down"—everything was up, sideways, or floating freely! They passed a large window that looked out into space, and Luna's breath caught in her throat. She could see the entire Earth below—its swirling clouds, the deep blue oceans, and the green and brown landmasses.

The Earth looked so peaceful and beautiful from space, like a giant marble floating in the darkness.

"This is one of the best parts of living in space," Commander Sarah said, noticing Luna's gaze. "We get to look out these windows every day. It's a reminder of how special Earth is."

Luna nodded, feeling a deep connection to the planet below. She couldn't believe she was actually here, looking out at the world she called home from so far away.

Fun Fact: The space station is like a big science lab in space. Astronauts live and work there for months at a time, conducting experiments to learn about space, science, and how humans can live in space. Some experiments include studying how plants grow in microgravity, how muscles and bones change in space, and how materials behave in the absence of gravity.

As they continued the tour, Luna couldn't help but think about all the exciting discoveries that were happening in the space station.

She knew that each experiment, each new lesson, would help humans understand more about space —and maybe, one day, help people travel to faraway planets like Mars.

Luna's heart swelled with pride as she realized she was part of something much bigger than herself. She was standing in the place where astronauts made history every day—and she couldn't wait to see what new adventure awaited her next!

Chapter 5:
The Amazing Solar System

After Luna had explored the space station and floated in weightlessness, Commander Sarah invited her to learn more about the incredible planets in our solar system. Luna couldn't wait to hear more about the vast, mysterious world beyond the space station.

"Do you want to take a tour of the solar system?" Commander Sarah asked, with a wink.

Luna's eyes lit up. "You mean, we can actually visit the planets?"

Commander Sarah smiled. "Well, not quite. But we can learn all about them with the help of some amazing technology!"

They entered a special room in the space station, filled with large screens and interactive models of the solar system. Commander Sarah tapped a few buttons, and suddenly the room lit up with pictures and animations of planets, moons, and asteroids.

"This is our solar system," Commander Sarah said, pointing to a glowing model of the Sun at the center. "The Sun is the star that gives us light and warmth, and all the planets orbit around it. The solar system is made up of the Sun, the eight planets, their moons, asteroids, and comets." Luna stared at the bright model, amazed at how everything in the solar system seemed to fit together perfectly. "Wow, it's like a giant cosmic dance!" she said, eyes wide.

"Exactly," Commander Sarah agreed. "The planets move around the Sun in a path called an orbit. Each planet has its own unique characteristics."

Luna's curiosity grew even more. "Tell me about the planets!"

Commander Sarah tapped another button, and a hologram of the first planet appeared: Mercury.

"Mercury is the closest planet to the Sun. It's tiny and super hot because it's so close to the Sun.

Even though it's small, it has no atmosphere to keep the heat in, so it can get really cold at night!"

Luna imagined a planet so hot you could fry an egg on it. It sounded strange, but fascinating.

Next, a picture of Venus appeared. "Venus is often called Earth's twin because it's similar in size. But it's much hotter! Venus has thick clouds that trap heat, making it the hottest planet in the solar system."

Luna wrinkled her nose. "That doesn't sound like a fun place to visit!"

Commander Sarah chuckled. “No, not really. But it’s interesting to study.”

Then, a beautiful image of Earth filled the screen. Luna smiled as she looked at the swirling oceans and landmasses.

“Earth is the only planet known to have life,” Commander Sarah said. “It’s a special place, with the perfect conditions for plants, animals, and humans to live. It’s the only planet with liquid water on its surface.”

“It’s amazing,” Luna whispered, feeling proud of her home planet.

The next planet, Mars, appeared on the screen. Luna's eyes sparkled.
"Mars is sometimes called the 'Red Planet' because of its reddish soil. It has huge volcanoes, deep canyons, and a very thin atmosphere. It's too cold for humans to live there without special equipment, but scientists believe that someday, humans might land on Mars!"
"Someday, I want to be one of those humans!" Luna exclaimed, her dream of visiting Mars stronger than ever.

"We're getting closer," Commander Sarah said. "With the technology we have today, we're learning more and more about Mars. There are even rovers, like the Perseverance Rover, that are already exploring its surface. You'll be able to see it up close one day!"

Then, the screen zoomed out, showing Jupiter, the largest planet in the solar system.

"Jupiter is a giant gas planet.

It's so big that you could fit over 1,300 Earths inside it! Jupiter has a strong magnetic field and more than 70 moons. And, of course, it has the Great Red Spot, a massive storm that's been raging for hundreds of years."

Luna's mouth dropped open. "That's HUGE! I can't even imagine a storm that big!"

"It's incredible," Commander Sarah said, her eyes twinkling. "Jupiter's moons are also fascinating, with some even having their own oceans under the surface. Scientists are studying them to see if they could support life!"

Next came Saturn, a planet Luna had always found beautiful because of its stunning rings.

“Saturn is known for its beautiful rings, made of ice and rock. These rings stretch out for thousands of miles! Saturn is another gas giant, like Jupiter, but it’s known for its rings, which are the most impressive in the entire solar system.”

Luna pressed her face against the screen, imagining how incredible it must be to see Saturn in person. “Wow, its rings are like a giant cosmic necklace!”

"Exactly," Commander Sarah said. "Saturn is beautiful, and its rings are just one of the many wonders of our solar system."

Next came Uranus and Neptune, both icy blue planets far from the Sun. Commander Sarah explained that they were both made of gas and ice, and their extreme distance from the Sun made them freezing cold.

Luna was amazed by the diversity of the planets.

Each one had its own unique characteristics, and they all orbited around the Sun in perfect harmony.

Fun Fact: The solar system is made up of the Sun, all the planets, moons, asteroids, and comets that orbit it. Mars, known as the "Red Planet," is named for its reddish color, caused by iron oxide (rust) on its surface. It's the fourth planet from the Sun and has long been a target for exploration. Scientists hope that someday, humans will land on Mars to learn more about its surface and maybe even build a base there!

"There's so much to learn about our solar system," Luna said, her eyes full of wonder. "I can't wait to explore more one day!"

"You're already off to a great start, Luna," Commander Sarah said with a smile. "Who knows? Maybe one day you'll be the one exploring Mars in person."

Luna's heart soared as she thought about the possibilities. With so much to explore and discover, she knew her adventure in space was just beginning.

Chapter 6: The Return to Earth

After an exciting day exploring the space station and learning about the wonders of the solar system, Luna knew it was time to say goodbye. She had experienced more in one day than most people could ever dream of, and her heart felt full of memories that would last a lifetime.

“I can’t believe how much I’ve learned!” Luna said, floating over to Commander Sarah and Captain Liam.

"It was an honor to have you here, Luna," Captain Liam said, giving her a friendly pat on the back. "You've been an excellent space explorer."

"You're welcome to visit again anytime," Commander Sarah added with a smile. "We're always learning new things up here, and we'd love to share them with you."

Luna smiled, feeling a little sad that her adventure was coming to an end, but she also felt excited.

She had so many stories to tell her friends, and she couldn't wait to share everything she had seen and learned.

As Luna floated over to the shuttle docking area, she turned back to look out the window one last time. The Earth below was beautiful, with its swirling clouds and vast oceans. Luna felt a deep connection to it, knowing that she would always carry a piece of space in her heart.

"Goodbye, space," Luna whispered, her voice full of awe. "I'll be back one day."

The shuttle ride back to Earth was an adventure in itself. Luna strapped herself into her seat and looked out the window. The stars were twinkling, and she could still see the glowing space station off in the distance.

As the shuttle began to power up for re-entry, Luna could feel the shuttle start to shake. “Here it comes,” she said, gripping the armrests tightly. The rocket’s engines roared, and the shuttle began to plunge back toward Earth. Luna could feel a little rumble as they passed through the thick atmosphere.

It was bumpy, like being on a roller coaster, but she wasn't scared—she was too excited to be home!

"Hold on tight!" Captain Liam called from the front of the shuttle, making sure everyone was safe. "We're going to re-enter the atmosphere now, so things will get a little bumpy."

Luna squeezed her eyes shut for a moment as the shuttle hit the thick air and began to slow down. The heat shield on the shuttle's underside protected it from the intense heat as they entered the atmosphere at high speed.

Luna felt like she was flying faster than ever, but she knew the shuttle was built to handle the heat and pressure. Suddenly, the shaking began to subside. Luna opened her eyes and saw the blue sky of Earth below. The shuttle had slowed down, and everything was calm again.

“We’re almost there!” Commander Sarah said, her voice steady and calm.

Luna looked out the window and saw the familiar patchwork of fields and forests below, slowly growing larger.

The shuttle's landing gear extended, and the wheels touched down softly on the runway. Luna could feel the shuttle come to a gentle stop, and she knew she was safely back on Earth.

"We did it!" Luna exclaimed, clapping her hands. "We made it back!"

The astronauts laughed and unbuckled their seatbelts.

"Welcome home, Luna!" Captain Liam said with a grin.

Luna took one last look at the sky, knowing that the stars above would always remind her of the incredible journey she had just experienced.

"Thank you for everything," she said to her new astronaut friends. "I'll never forget this." Fun Fact: Space shuttles land just like airplanes. After astronauts have completed their missions, they carefully re-enter the Earth's atmosphere. The process of re-entry is tricky, because the shuttle must slow down without burning up from the intense heat of the atmosphere. Special heat shields protect the shuttle during this phase, keeping astronauts safe as they return home.

As Luna stepped out of the shuttle and onto the ground, she couldn't wait to share her space adventure with her friends and family. She knew they would be amazed by the stories she had to tell, and she couldn't wait to tell them all about the planets, the space station, and the weightlessness of floating in space.

But Luna also knew that this adventure wasn't the end. One day, she hoped to return to space.

Maybe, just maybe, she would even be one of the astronauts who helped make history by landing on Mars.

For now, she looked up at the stars one more time.

“Goodbye, space,” Luna whispered. “Until next time.”

And with that, Luna’s incredible adventure in space came to an end—but the dream of returning one day would live on in her heart forever.

Luna has returned to Earth, but her journey has only just begun. With her heart filled with wonder and her mind buzzing with the knowledge of the cosmos, she dreams of what lies beyond. Luna realizes that the universe is vast and full of mysteries, and she is eager to explore them all. Her time in space has taught her valuable lessons about the planets, stars, and galaxies, and she now understands that the adventure of discovery is endless.

With her passion for space ignited, Luna starts planning her next grand journey. This time, she dreams of traveling to the Moon, a place she's longed to explore ever since her first adventure. She imagines walking on its silvery surface, gazing at Earth from a distance, and uncovering secrets hidden for billions of years. But Luna's ambitions don't stop there. She dreams of venturing even further – to distant stars and unknown planets. Perhaps one day, she'll even discover life on a faraway world, or witness a supernova explosion from the safety of a spacecraft.

As Luna prepares for her next great adventure, she knows she will face challenges along the way. There will be new obstacles to overcome and more mysteries to solve, but with her curiosity and determination, she is ready for whatever comes next. She understands that the universe is a puzzle that may never be fully solved, but she's excited to try.

Fun Fact: The universe is full of unsolved mysteries, and scientists are constantly discovering new things about space.

Every day, they learn more about distant planets, black holes, and the origins of stars. Who knows? Maybe one day, Luna will be one of those scientists, uncovering the secrets of the cosmos and helping humanity understand the wonders of the universe. After all, with her adventurous spirit, there's no limit to how far she can go!

www.ingramcontent.com/pod-product-compliance
Lightning Source LLC
LaVergne TN
LVHW010502160826
845677LV00012B/2615

* 9 7 9 8 2 2 4 1 5 5 3 8 5 *